I0074489

HOW DO BEES (AND HUMANS) SEE GREY LEVELS?
Adrian Horridge

ISBN: 978-1-914934-54-4

Text and graphics: Adrian Horridge. www.adrian-horridge.org.

Published by Northern Bee Books 2023
Northern Bee Books, Scout Bottom Farm, Mytholmroyd,
Hebden Bridge HX7 5JS (UK).
www.northernbeebooks.co.uk
+44 (0) 1422 882751.

Book design by www.SiPat.co.uk

HOW DO BEES (AND HUMANS) SEE GREY LEVELS?

Intuitive, logical, and anti-intuitive explanations for some curious observations

Adrian Horridge

HOW DO BEES (AND HUMANS) SEE GREY LEVELS?

Contents

What's it all about?

Since the nineteenth century there have been many descriptions of bees' ability to discriminate coloured papers, but it is less well known that they can be trained to come to a grey target showing no colour. From their recognition of grey, and ability to discriminate light grey from dark grey, and from white and black, von Frisch concluded that bees have a form of achromatic vision in addition to well-developed colour vision.

The problem is that there are no grey or black photons, so how do bees or humans see grey? The unsupported assumptions of von Frisch, the sequence of his conclusions, and his fatal errors, will be examined, and right thinking restored.

This booklet is a supplement to my recent book, 'The Discovery of a Visual System: The Honeybee.' (Horridge, 2019). In another recent publication, bees were trained to come to a plain grey target, and some bees were reared seeing only grey feeders. It was thought that training on grey would remove any previously acquired preferences for a colour (Hempel di Ibarra et al, 2022). Without further training, they were tested with a variety of colours. Indeed, the bees responded when presented with variously coloured new targets. However, the conclusion that they had learned the grey and then responded to various colours in meaningful ways, is contrary to all that we know about light, colour, black and bee vision.

I hope to present an interpretation of bee and human responses to grey, and then to various colours, based on simple physical principles.

The establishment of belief by Karl von Frisch

Humans very readily distinguish patterns of black on white paper, as the reading of print demonstrates. They also distinguish grey shades from each other, and colours from all shades of grey. For humans, this has long been a standard test for colour vision.

In 1914, Karl von Frisch had been using this test on a variety of fishes at the Naples Marine Laboratory. He believed that bees had human-like vision of colours and grey levels, and set to work to prove it with the test designed for human vision. He purchased a standard set of 15 different grey levels ranging from white to black. Each paper was a square of 15cm each side, so that bees could easily find them (von Frisch, 1914). He was unaware that these papers were made from wood pulp that did not reflect ultraviolet, with the result that the bees' UV receptors were excluded from the tests. He joined together the whole set of 15 grey papers, more or less at random, together with the training colour, to form a 4x4 display (Figure 1). This test card, 60 × 60 cm, was presented to each trained group of bees.

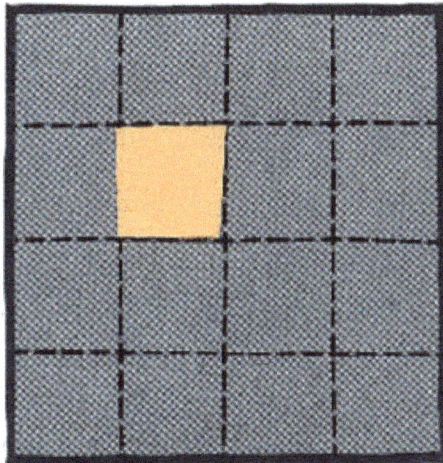

Figure 1. von Frisch packed together 15 different grey levels and one colour on a test pad that was presented to each group of trained bees.

The results were promising but mixed. Bees trained on a blue paper usually preferred blue to all shades of grey. Those trained on yellow went to yellow rather than a shade of grey. Karl noticed that they were always attracted to edges of the targets, but he did not make use of this vital fact. Bees trained on extremes of light or dark grey could usually select a similar grey level if the choice was limited. He assumed that bees see grey and black, and called it "achromatic vision".

Von Frisch found that yellows and blues at each end of the colour series were easily recognised, but he tried five types of green with no success. For example, bees trained on blue-green could not find the training paper among grey papers and could not discriminate it from other colours. Strangely, bees trained on mid-grey could not distinguish mid-grey when it was presented in the whole grey series. Von Frisch could not explain, and gave up on training on grey. He did not realise that bees could not distinguish such small differences in brightness, but they used an average value, equal to mid-grey, for the whole set on the card.

Because there were many successes, *he ignores the anomalies*, and publishes his conclusion that bees distinguish colours and grey levels, and therefore they have colour vision similar to that in man, and that besides colour, they had a kind of achromatic vision.

Von Frisch published 180 pages of his observations in German all together in a rather obscure journal, with an unusual note to say that copying was forbidden and translation was reserved, which perhaps explains why the original remained little known or discussed, and seems to suggest that curious investigators were not welcome. Subsequently, colour vision of bees was universally accepted in text books, except that in 1918, Professor Hess of Ophthalmology in Münich, pointed out that bees trained on a chequerboard of blue and yellow squares easily learned to go to blue, but they learned nothing when rewarded on the yellow squares; but he also had no explanation. Hess died in1923 and was soon forgotten, and von Frisch and his students published many later papers in his own journal, always accepting bee discrimination of colour, black, and grey levels.

Bees could not distinguish the grey levels

Later, von Frisch must have been aware of a publication by Ernest Wolf, a bee trainer who left Germany with his wife Gertrud Zerrahn before 1933 to work with Selig Hecht, at Columbia University. This was a formidable group that studied the physics of vision in man and many animals, for which Hecht was awarded a Nobel Prize. Wolf measured the resolution of the bees' detection of changes in brightness. Surprisingly, for bees, the minimum detected change was 25% in bright light and 300% in light two log units less bright (one hundredth). Corresponding values for the human eye are a 2% difference in bright light and 20% at 3 log units down (See my book, Horridge (2019), p. 17), and Wolf (1933).

These results show that bees could not possibly detect the differences between fifteen adjacent grey levels in the commercial series of papers. Presumably, on average about 7% difference in brightness between adjacent papers was normally used for testing colour vision in man. In the group of 15 or 16 test papers joined together, bees would not detect all the boundaries between different grey levels, so they sum together the whole lot as one area of average grey. No wonder there was a good deal of unexplained variability in von Frisch's results; his bees could discriminate only some of his papers.

This fatal flaw at the heart of von Frisch's test makes nonsense of his whole effort. Where they cannot discriminate the boundary between two test papers, bees fuse the areas together as one, and take an average value of the green contrast at boundaries. In fact, Frisch says that mid-grey was indistinguishable from the whole test. His handy 4x4 test display (Figure 1) had ruined his results.

Anomalies were slowly noticed

At the end of the last century, I and several of my students working with bee vision, accepted that bees had some sort of colour vision, and anomalies were not prominent in my first book (Horridge, 2009). We discovered, for example, a movement of black on white in the vertical direction was actually detected by the opposite change in position of the white in the background (actually the blue content of the white). Anomalies published by Ronacher (1979) and Giurfa and his students (1996, 1998), that were difficult to explain by von Frisch theory, were largely ignored. When I turned to colour vision in 2012, I soon found many anomalies, and began a long series of new experiments to distinguish the feature detectors of the honeybee for colours and grey levels.

First, I had realized from the observation of von Frisch, that the eye of the bee measures the sum or average of the stimulus over the area enclosed by a boundary of green contrast, even, over the whole eye. This explained why his bees could not distinguish a mid-grey paper when tested against the 15 grey levels presented together (Figure 1). The bees had shown that the average grey level of the 15 test papers presented together was the same as that of the mid-grey. Actually, the bees did not even have available the grey levels of single test papers. They behaved as typical dichromats, and detected only the average level of blue in the white content of grey areas, and the total of green contrast at edges.

Von Frisch had no idea of the properties of the individual receptors or number of types in the honeybee compound eye, so he could believe in their achromatic and colour vision; in fact, he could believe anything, and make his followers comply. Later the spectral sensitivity curves of the photoreceptors were carefully measured by Autrum and von Zwehl (1964) (Figure 2).

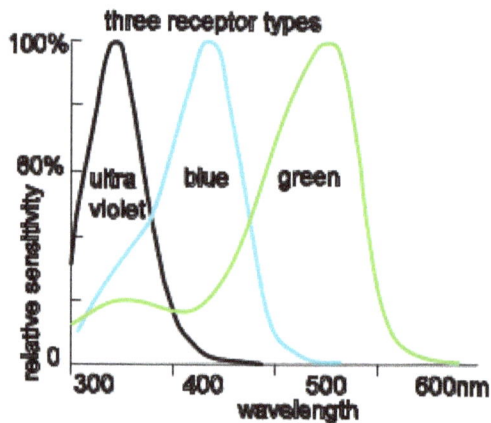

Figure 2. The spectral sensitivity of the three types of receptor cell in each ommatidium of the honeybee. The curves have been normalized to the same peak of 100%, but in reality, the UV receptors are at least ten times more sensitive than the others, and have a different function, to detect the direction of the sky as an escape route.

Honeybees have three types of photoreceptor cells in each ommatidium. Six of the eight cells have a spectral sensitivity peak in the green near 550 nm, one has a peak in the blue near 430 nm, and one cell in the ultra-violet near 350 nm. Natural foliage reflects little UV, but the UV cells inform the bee of the direction of the escape route to the UV in the sky, *and are not known to be connected to memory, or used in foraging.* Furthermore, in experiments, we can rule out the UV receptors because the test papers did not reflect UV. As a result, **the bees were functionally dichromats**.

Having measured the spectral sensitivity of the photodetectors, and the corresponding emission spectrum of all the experimental papers used, it is easy to calculate the relative total stimulus from any test paper to the blue and green types of receptor cells in the eye of the bee. That is nothing to do with bee behaviour; it is just physics of pigments and light. In my analysis of bee colour vision (Horridge, 2019) there are many examples of grey used as a colour, like any other colour, and this was accepted by the bees, but they had responded to the blue in the white, and green contrast at edges.

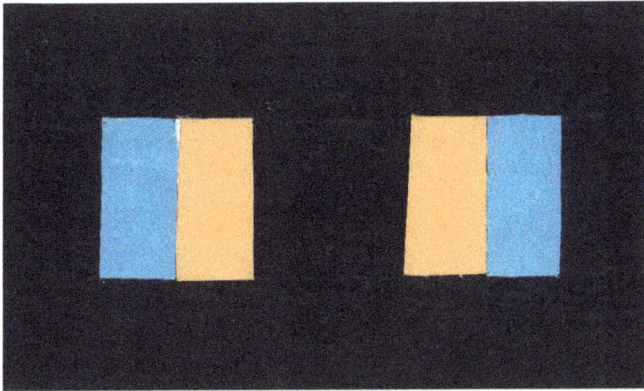

Figure 3. Patterns of two colours with no green contrast at the central boundary. Bees are unable to discriminate these mirror images because they detect no separation of colours

Tests for functional dichromats

We can make patterns of two colours that stimulate equally either green or blue receptors of bees. They are called equiluminant colours, and a pattern combining them alone cannot be distinguished from its mirror image (Figure 3). However, if a vertical black line, or any pattern of green contrast, is added to an equiluminant pattern (Figure 4), its polarity is distinguished immediately because the two colours are processed separately.

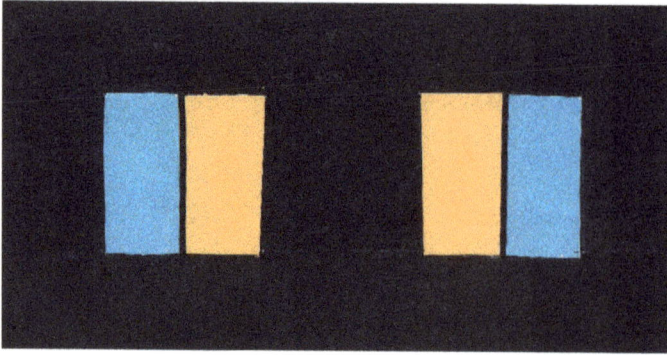

Figure 4. When a vertical black bar separates the colours, they are detected separately, and the bees discriminate these mirror images.

The basic mechanisms for seeing

Bees usually scan as they search while feeding. Behind the retina are a few simple feature detectors that each respond in a different way to two aspects of the visual inputs via the photoreceptors, the total local amount of blue, and the total local modulation of the green receptors. Modulation is the length of edge multiplied by the green contrast at each part of the edge. The combination of feature detectors tells the bee the polarity (Figure 5) and provides enough combination of responses to learn and remember the location.

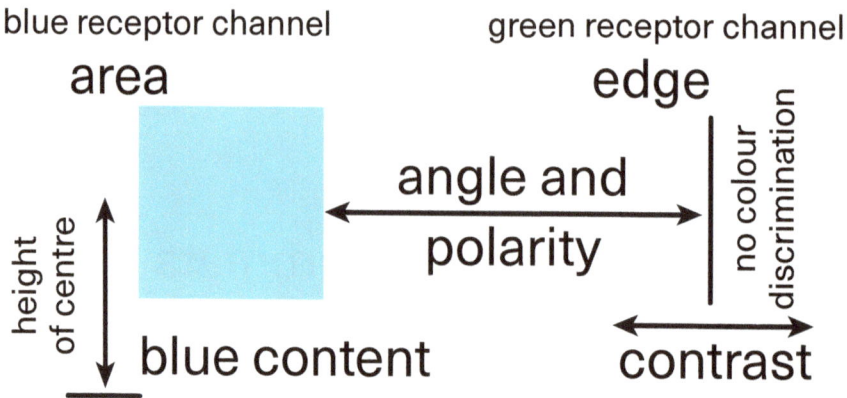

Figure 5. The polarity of any pattern can be detected, learned, and remembered when it displays green contrast on the left or on the right of a patch that emits blue. This is one of the one of the ways that the two inputs of a dichromat normally collaborate when following a feeding route.

Left-right polarity is detected and learned

Bees flying to a target are not restricted to the recognition of places. When a target displays green contrast at on side of a patch that excites blue receptors (Figure 5), or when there is a left-right gradient of green contrast (Figure 6), bees are very sensitive to the left-right polarity. They can use polarity as signpost to turn left or right. The actual colours or patterns are irrelevant; the cue is the left-right polarity.

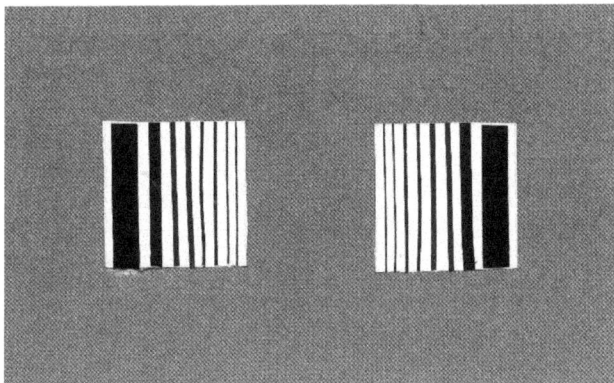

Figure 6. The polarity of a gradient of modulation is detected, and can also be used like a sign post, indicating a turn to right or left.

Grey is not special for bees

Bees detect and remember colours by their blue content and by green contrast at edges, especially edges of shadows. These variables can vary independently, making many discriminations possible. Bees detect grey like any other colour, by green contrast and content of blue. Grey, however, is a colour in which the blue content and green contrast at edges are both locked to the percentage of white. Each grey level is fixed in its content of blue.

It was a mistake to train dichromatic bees on grey and then test them with flowers looking for evidence of colour preferences. Some were even tested for preference for chromatic contrast, which assumes full trichromatic colour vision. The trained bees would simply look for resemblance to blue content in the grey, or green contrast at edges of grey.

There are no grey photons

Let us now consider the fundamental properties of grey, which is said to be a mixture of white and black, or a shade of white or a tint of black. You can make a well-known grey powder by mixing together finely ground black charcoal with powdered white potassium nitrate. Gunpowder certainly looks grey, but the difficulty is that you cannot have grey rays of light emerging from it and reaching your eye, because there are no grey or black photons coming from the charcoal.

Now stare hard at a grey surface, preferably one that us dimly illuminated. Eventually you can usually see with your naked eye that it consists of millions of tiny black dots on a lighter background. You never *see grey*. As a grey surface darkens towards black, you detect a thinning out of the white photons, and hallucinate more and more tiny black dots, so you hallucinate the colour grey in your brain. The black dots are gaps in the smooth flow of white photons. They stay in focus and the same size when you examine them with a lens. I see them when I look at a bright object with my eyes shut, therefore they are inside my eyes. Apparently you hallucinate grey when you look at white but there are insufficient photons. The black dots reveal how we construct the graded nature of our hallucination of grey. Black-white photographs also have only various densities of black silver grains.

How do bees detect grey?

Now consider a bee looking at grey. The white content of grey can be between 0 and 100%, and white contains about 50% blue, and a bee would detect the blue content. At the edges of grey, a bee would detect green contrast, especially at edges where shadows generate strong signals. Grey as a colour, therefore, is detected by the honeybee by the blue content of the white component in the grey, and by green contrast at its edges, --- no different from any other colour as seen by a dichromat. Grey as a colour has one special feature however; there is only one measure, the white/black ratio, which fixes both the green contrast and the blue content together.

Do bees see grey levels or black? No! They detect blue, blue contrast, and especially green contrast. This dichromatic vision is common to many insects. Cats and dogs, ungulates, and most eutherian mammals have this type of vision, which has evolved on a planet where every panorama is composed of colours with more or less blue than the universal background of green chlorophyl. Even yellow contains about 10% of maximum stimulus to the blue receptors. Changes in position of green contrast at edges of shadows are detected as motion, and then further processed in local regions to detect relative motion and parallax. There is nothing special about grey for a dichromat.

Bees use their vision like quantity surveyors

A bee that sees a fence of vertical posts can learn and remember only the *average orientation* of edges and the *amount of modulation*, which is the length of edge, multiplied by the contrast at each bit of edge. Similarly, an array of spots is summed and only the total modulation learned. A mixture of edges at angles to each other is summed, so that the orientations maybe are all lost. These are example of reduction of the information load, without loss of some recognition features. Relative movement is also detected to measure parallax and to avoid a crash.

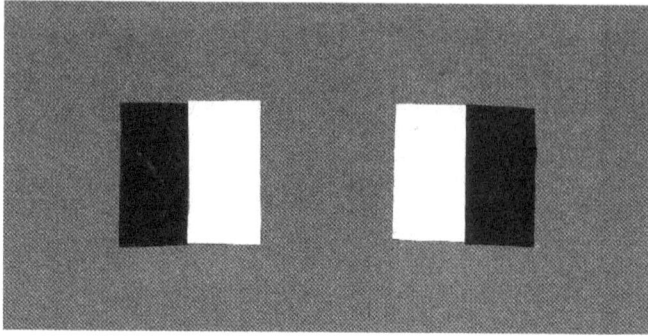

Figure 7. A black square next to a white one is not discriminated from its mirror image, because the bees are unaware which way the vertical contrasts were scanned.

Bees usually discriminate mirror images except in some instructive examples. We discovered long ago that a black square at the side of a white square, on a grey background (Figure 7), is not discriminated from its mirror image, but a black square above a white one is readily distinguished from the reverse (Figure 8). The explanation is that bees continually scan in the horizontal plane, they detect a part of the stream of blue and green modulation, and remember the total for each scan. At any one time the visual flow from a scan is not re-assembled to give detailed spatial information about relative locations. The totals are sufficient, though they would be confusing in human vision. The bee visual system detects and learns totals of blue and of modulation to recognise a place, and ignores how local detail is spaced during the scan in the horizontal plane. There is an essential reduction of the information load, but there is a price to pay, loss of the spatial relations.

This local summation and memory of the totals gives the impression that bees can count, but detailed tests of trained bees show that they have learned relative amounts of edge modulation, not absolute numbers.

As said, bees fail to discriminate a horizontal combination of a black and a white square from its mirror image, so it is probable that they do not distinguish a horizontal combination of grey levels from its mirror image, even when there is more than 30% of intensity difference. However, they distinguish very well when the modulations are at different heights (Figure 8).

Figure 8. Black squares and white squares at different heights are readily discriminated.

Two further details require care when training and testing bees to choose between two targets. Bees usually learn first the unrewarded pattern because that is where they have to make an active response and change their preference. They learn nothing when they first visit the rewarded target.

Secondly, when trained on a symmetrical pattern versus an unsymmetrical one, they detect and learn the asymmetrical one because they have detectors of left-right polarity. They never learn the symmetrical target first, because they have no feature detectors for symmetry. In fact, there can be no general detector of symmetry in all its forms.

Detection of motion

All motion-detector neurons in the visual systems of insects, and probably of most animals, have input only from green receptors. They have evolved in this way because motion must be detect by only one channel to avoid interference, and green is the commonest colour in the panorama and also has the most useful total energy in the illumination by the sun. Intensity of green is not a useful measure because green is everywhere, and *this channel is used by all dichromats to detect green contrast, and changes of green intensity, not the intensity itself.* This makes green receptors most sensitive to edge movement, relative movement and parallax.

The progressive selection towards detection of smaller and faster green contrast, has carried with it the improvement in spatial resolution and sensitivity of the whole insect visual system towards the physical limits. This is true for all dichromats; just two visual pigments for vision, green contrast for motion and acuity, percentage of blue content for the place on the colour scale. In agreement with this, all green sensitive neurons in the optic lobes and brain of the bee respond to changes in intensity; none measure intensity.

The UV receptors of the bee

The UV receptors in every ommatidium of the bee are not useful for colour vision because most natural objects reflect little UV. Also, the UV cells are at least 10 times more sensitive than the green and blue cells, as if they must operate down to low light levels. Bees in the escape response detect the direction of the sky when disturbed. So, for most seeing, the bee is effectively a typical dichromat, with one input measuring amount of blue, or blue contrast, and the other input the green contrast at edges. This simple combination for seeing surfaces is the result of long selection for sensitivity, speed of action, best resolution, low noise, and eye size, on a planet where the strongest part of the sunlight spectrum is in the green, most stuff is green, and items of interest like flowers or predators are less blue, or more blue, than the green background. All animal visual systems have evolved in the living world, green with chlorophyl; UV only indicates the direction of the sky----up, and is of little use unless you fly.

Mammals are also commonly dichromats

By simply looking at sections of frozen retina with a powerful microscope it is possible to measure the absorption spectrum of the individual cones in the eyes of most animals. Not surprisingly, because they have evolved in the same natural world, cats, dogs, ruminants, and most common diurnal mammals, are dichromats with only blue cones and green cones. The optics of their eyes also have an F number between 2 and 10, because cameras and eyes evolved in the same light intensities. Mammals make better use of the signals than bees because they integrate inputs of neighbouring receptors and build up small regions to make a panorama; bees have only a few feature detector responses. Dichromacy performs well without a large cerebral cortex, but clever trichromats, especially primates, require a large cortex which is almost entirely taken up by vision.

Eyes evolved as an aid for walking or flying, not for the recognition of food or a mate, which is always more reliably done by odour. Owners of a dog or cat usually will tell you that their pets see only black, white and grey levels, but these animals also evolved in a world of green with items of blue or yellow, not in a world of black and white. All are dichromats but use odour receptors for final recognition, because they are more certain, and are far more numerous.

Incorporating a temporary memory of vertical edges

When a bee, dog or crab, or other animal is placed at the centre of a drum or simply sees the natural world, it builds up a spatial memory that is easily demonstrated. The lights are all turned out, and while the animal is briefly in the dark, everything is rotated a few degrees around it. When the light returns, the animal, or just its eyes, rotate to bring the panorama back to the same relative position as before. The visual system enables an animal to remember briefly where things were located relative to each other before the eyes moved, and so detect the shift. From the shift in position, called parallax, they obtain spatial relationships in three dimensions.

Humans really are special

Today, billions of people read black print on white paper without a thought or error, with their trichromatic vision and a huge cerebral cortex. They are totally unaware of a crucial first step, which is to compensate for the chromatic aberrations caused by the lenses in their own eyes, which are continually being corrected. Also, disruptions of the image, the effect of the blind spot, and mismatch between the two eyes, are totally wiped away by mysterious compensation processes in the visual cortex.

Humans see the brightest object in the visual field as white, even white leaves among green foliage in sunlight. We can look at pure black, and believe we see black, but we simply hallucinate black where there are no photons, and similarly we hallucinate grey where photons are fewer than required to see white.

Good intentions but inadequate data

'The Royal Society's motto 'Nullius in verba' is taken to mean 'take nobody's word for it'. It is an expression of the determination of Fellows to withstand hearsay and guesswork, and requested experimental proof. Accordingly, the work of von Frisch was his idea of experimental proof that bees see colours as humans do. The topic went astray because at the time there was insufficient basic knowledge to design crucial experiments. No-one could think of a better theory, but now we have a theory that has been thoroughly tested by calibrating the colours and receptor cells by physical methods, besides vigorous testing with many possible features for detection by the bee. The results are undeniable

Little progress

There are still followers of von Frisch, building on his conclusions. It is unbelievable, but they publish in the best refereed journals. They get past the referees because an organized group of them recommend each other as referees, and suggest no honest ones, so unsuspecting editors are duped. They are easy to recognise because without further tests, they assume that bees learn the item they have been trained on, whereas bees have a few types of feature detectors, and many tests are necessary to identify what they have actually detected in the targets. An example was in 'Biology Letters doi 10.1098/rsbl.2010.1056 22 Dec, 2010'. The referees there included the same closed referee circle as in the recent example in Phil. Trans Roy Soc B. Little could be done, because when I complained, the Royal Society put a temporary block on my email.

References

Autrum, H. and von Swehl, V. 1964 Spektrale Empfindlichkeit einzelner Sehzellen des Bienenauges. Zeitschrift für vergleichende Physiologie. 48, 357-384.

Bowmaker, J. K. 2013 Evolution of the Vertebrate Eye. Chapter 23 in "How Animals See the World." Eds. Lazareva, O.F., Shimizu, T. and Wasserman, E.A. Oxford University Press. 2012

Frisch, K, von, 1914 Der Farbensinn und Formensinn der Bienen. Zoologische Jahrbucher. Abteilung für algemeine Zoologie und Physiologie der Thiere. 15, 1-188.

Giurfa, M.,Vorobyev, P.,Brandt, R., Posner, B. and Menzel, R. 1997 Discrimination of coloured stimuli by honeybees: alternative use of achromatic and chromatic signals. Journal of Comparative Physiology 180 235-241.

Hempel de Ibarra, N., Holtze, S., Bäucker, C., Sprau, P. and Vorobyev, M. 2022 The role of colour patterns for the recognition of flowers. Phil. Trans. Roy Soc B. 377, 20210284.

Horridge, G. A. 2009 *What Does the Honeybee See? And How Do We Know*. ANU E Press. Canberra. 360 pp.

Horridge G. A. 2019 *The Discovery of a Visual System: The Honeybee*. CABI Books. Boston and Wallingford. Pp. 274

Ronacher, B. 1998 How do bees learn and recognise visual patterns? Biological Kybernetics 79, 477-485.

Wolf, E. 1933 The visual intensity discrimination of the honeybee. Journal of General Physiology. 16, 407-422.

About the author

Born 1927, after broken wartime schooling, then 4 years as a scholar at St John's College, Cambridge, Adrian spent 10 years working out details of nervous systems and nervous control of movement in the phylum Coelenterata, including jellyfish, medusae, and corals, and also Ctenophores. There followed a year at the Center for Study of the Behavioral Sciences at Palo Alto, with Ted Bullock, writing a huge two-volume work on *Invertebrate Nervous Systems* (1965).

When Adrian returned to St. Andrews, Scotland, in 1961, he had decided to concentrate on all aspects of the arthropod compound eye. His first students recorded from the photoreceptors, and described the basic optics of the locust and fly eyes and their capture of single photons at extremely low light levels. He subsequently

Examining Sweet Autumn Clematis. Bees see white mainly with their blue receptors.

published about 250 papers on the compound and optic lobes of many kinds of insects. In 1969 he was elected to the Royal Society and in that year became one of four Founder Professors of Biological Sciences in the Australian National University where the work on insect vision continued. In 1990, discoveries of how insects pilot themselves in flight, together with his early experience doing National Service as an aircraft structural engineer at Farnborough, led to applications in drone helicopters and planes, with computer and vision on board, supported by American funds. For more detail, see his web page at Adrian-Horridge.org

Adrian retired in 1992 and turned to the analysis of what bees see, by using a Y-choice maze. This analysis led to the discovery of feature detectors for combinations of edges, circles and spokes in one channel for green contrast at edges, and for content and height of blue in a separate channel. Left/right polarity between an area of blue and of green contrast was also detected, learned, and used as a signpost. The UV receptor in each ommatidium seems to be used only for registering the direction of the sky as an escape route. Hundreds of hours training bees and testing

Adrian with grandson Silvan and a Valota speciosa plant in flower. Bees see red like black, mainly by the green contrast at edges.

them for what they had learned led to a book "*The Discovery of a Visual System: The Honeybee*" (2019).

Adrian is also known for his books and papers which record the lost maritime ethnology of Indonesia. In the 1970's and 80's he made a unique photographic record of many local types on many distant islands. That World Heritage fleet of thousands of fishing boats and outrigger canoes has now become motorised, designs changed, and rigging gone. There is little improvement in the catch because there was always overfishing and depletion, and now there is carbon dioxide emission from the engines and expenses of fuel and maintenance.

HOW DO BEES (AND HUMANS) SEE GREY LEVELS?

www.ingramcontent.com/pod-product-compliance
Lightning Source LLC
Chambersburg PA
CBHW040154200326
41520CB00028B/7597